The Global Technology Revolution 2020

EXECUTIVE SUMMARY

Bio/Nano/Materials/Information Trends, Drivers, Barriers, and Social Implications

T0195503

Richard Silberglitt • Philip S. Antón • David R. Howell • Anny Wong

with S. R. Bohandy, Natalie Gassman, Brian A. Jackson, Eric Landree, Shari Lawrence Pfleeger, Elaine M. Newton, and Felicia Wu

Prepared for the
National Intelligence Council

NATIONAL SECURITY RESEARCH DIVISION

The research described in this report was prepared for the National Intelligence Council.

Library of Congress Cataloging-in-Publication Data

The Global technology revolution 2020, executive summary : bio/nano/materials/information trends, drivers, barriers, and social implications / Richard Silberglitt ... [et al.].
     p. cm.
  "MG-475."
  Includes bibliographical references.
  ISBN 0-8330-3910-5 (pbk. : alk. paper)
  1. Nanotechnology. I. Silberglitt, R. S. (Richard S.).

  T174.7.G584 2006
  338.9'260905—dc22

                                                                    2006009729

The RAND Corporation is a nonprofit research organization providing objective analysis and effective solutions that address the challenges facing the public and private sectors around the world. RAND's publications do not necessarily reflect the opinions of its research clients and sponsors.

RAND® is a registered trademark.

Published 2006 by the RAND Corporation
1776 Main Street, P.O. Box 2138, Santa Monica, CA 90407-2138
1200 South Hayes Street, Arlington, VA 22202-5050
4570 Fifth Avenue, Suite 600, Pittsburgh, PA 15213
RAND URL: http://www.rand.org/
To order RAND documents or to obtain additional information, contact
Distribution Services: Telephone: (310) 451-7002;
Fax: (310) 451-6915; Email: order@rand.org

Foreword

The National Intelligence Council (NIC) sponsored this study by the RAND Corporation to inform the NIC's 2020 project[1] and help provide US policymakers with a view of how world developments could evolve, identifying opportunities and potentially negative developments that might warrant policy action. From June 2004 through August 2005, RAND undertook the challenging task of identifying technologies and applications that have the potential for significant and dominant global impacts by 2020.

As RAND found in its prior study for the NIC, *The Global Technology Revolution* (Antón, Silberglitt, and Schneider, 2001), technology will continue to accelerate and integrate developments from multiple scientific disciplines in a "convergence" that will have profound effects on society. RAND's new study, however, has delved further into social impacts and concluded that

- Regional and country-specific differences in social need and science and technology (S&T) capabilities are resulting in differences in how technology is revolutionizing human affairs around the world,
- Regional differences in public opinion and issues may strongly influence technology implementation,
- Maintaining S&T capacity requires consideration and action across a large number of social capabilities and stability dimensions,
- Capacity building is an essential component of development, and
- Public policy issues relating to some technology applications will engender strong public debate.

The implications of these findings are important to US policymakers. For example, while the United States remains a leader in S&T capability and innovation, it is not the sole leader and thus will not always dominate every technical area. Also, many technologies will evolve globally in ways that differ from their evolution in the United States, so we cannot merely apply a US view as a cookie cutter to understanding how technology will change the world. In addition, US understanding of potential technological threats from foreign powers requires a broad understanding not just of S&T skills and capabilities but also the institutional, human,

---

[1] See http://www.cia.gov/nic/NIC_2020_project.html for further information on the NIC 2020 Project and its final report, *Mapping the Global Future*.

and physical capacity to exploit technological opportunities. Finally, innovative combinations of new and existing technologies can help to meet region-specific needs despite their lack of use in the US sector.

I commend this report to you as a resource for understanding how S&T and social issues interact and depend not only on technological advances but also on the broader capabilities of countries that seek development and economic rewards through S&T exploitation. As important as S&T is today to the United States and the world, it will become even more important in the future.

Dr. Lawrence K. Gershwin
National Intelligence Officer for Science and Technology
Office of the Director of National Intelligence

Preface

Various technologies (including biotechnology, nanotechnology [broadly defined], materials technology, and information technology) have the potential for significant and dominant global impacts by 2020. This report is based on a set of *foresights* (not predictions or forecasts)[1] into global technology trends in biotechnology, nanotechnology, materials technology, and information technology and their implications for the world in the year 2020. These foresights were complemented by analysis of data on current and projected science and technology capabilities, drivers, and barriers in countries across the globe. For a more detailed discussion of the material described in this report, including further documentation and references, the reader is strongly recommended to review the in-depth analyses from this study.[2]

This work was sponsored by the National Intelligence Council (NIC) to inform its publication *Mapping the Global Future: Report of the National Intelligence Council's 2020 Project Based on Consultations with Nongovernmental Experts Around the World*, December 2004. In addition, funding was provided by the Intelligence Technology Innovation Center (ITIC) and the U.S. Department of Energy. It is a follow-on report to a RAND Corporation report, *The Global Technology Revolution* (Antón, Silberglitt, and Schneider, MR-1307-NIC, 2001), which was sponsored by the NIC to inform its 2000 document, *Global Trends 2015*. *Global Trends 2015* and the 1996 NIC document *Global Trends 2010* identified key factors that appeared poised to shape the world by 2015 and 2010, respectively.

This report should be of interest to policymakers, Intelligence Community analysts, technology developers, the public at large, and regional experts interested in potential global technology trends and their broader social effects.

This project was conducted jointly in the Intelligence Policy Center and the Acquisition and Technology Policy Center of the RAND National Security Research Division (NSRD). NSRD conducts research and analysis for the Office of the Secretary of Defense, the Joint Staff, the Unified Combatant Commands, the Department of the Navy, the Marine Corps, the defense agencies, and the Defense Intelligence Community, allied foreign governments, and foundations.

---

[1] A foresight activity examines trends and indicators of possible future developments without predicting or describing a single state or timeline and is thus distinct from a forecast or scenario development activity (Salo and Cuhls, 2003).

[2] See Silberglitt, Antón, Howell, and Wong (2006), available on the CD-ROM included with the hard copies of this report, or from the RAND Web site at http://www.rand.org/pubs/technical_reports/TR303/.

For further information regarding this report, contact its authors or the Intelligence Policy Center Director, John Parachini, at RAND Corporation, 1200 South Hayes Street, Arlington, VA 22202-5050; by telephone at 703.413.1100 x5579, or by email at john_parachini@rand.org. For more information on RAND's Acquisition and Technology Policy Center, contact the Director, Philip Antón. He can be reached by email at atpc-director@rand.org; by telephone at 310.393.0411, x7798; or by mail at RAND Corporation, 1776 Main Street, P.O. Box 2138, Santa Monica, CA 90407-2138. More information about RAND is available at www.rand.org.

Contents

Figures and Tables

Figures

Tables

Summary

This report presents the results from a set of foresights into global technology trends and their implications for the world in the year 2020. Areas of particular importance include biotechnology, nanotechnology, materials technology, and information technology. A sample of 29 countries across the spectrum of scientific advancement (low to high) was assessed with respect to the countries' ability to acquire and implement 16 key technology applications (e.g., cheap solar energy, rural wireless communications, genetically modified crops). The study's major conclusions include the following:

- Scientifically advanced countries, such as the United States, Germany, and Japan, will be able to implement all key technologies assessed.
- Countries that are not scientifically advanced will have to develop significant capacity and motivation before barriers to technology implementation can be overcome.
- Public policy issues in certain areas will engender public debate and strongly influence technology implementation.

Many technology trends and applications have substantial momentum behind them and will be the focus of continued research and development, consideration, market forces, and debate. These technologies will be applied in some guise or other, and the effects could be dramatic, including significant improvements in human lifespan, reshuffling of wealth, cultural amalgamation or innovation, and reduced privacy.

Acknowledgments

We would like to thank Lawrence K. Gershwin, Maj Gen Richard L. Engel (Ret.), William A. Anderson, Brian Shaw, and Julianne Chesky of the National Intelligence Council for their wonderful support and encouragement throughout this study.

The authors thank the following RAND regional experts for very helpful discussions of social and public policy issues, development needs, technological status, and the environment for implementation of technology applications: Keith Crane, Heather Gregg, Nina Hachigian, Rollie Lal, Kevin O'Brien, William Overholt, D.J. Peterson, Angel Rabasa, and Somi Seong. We also acknowledge the helpful discussions of quantum computing and cryptography we had with Calvin Shipbaugh and the several useful inputs on the status of science and technology in India from Ramesh Bapat, and are extremely grateful to Michael Tseng for quantifying the country data on capacity to acquire, drivers, and barriers.

The authors owe a special debt of gratitude to Robert Anderson, Steve Berner, Jennifer Brower, Ted Gordon, and Stephen Larrabee for their insightful reviews of this study and for several important suggestions that contributed greatly to improving the report. We also thank Linda Barron for her help in compiling, formatting, and producing the manuscript. Finally, we acknowledge the outstanding efforts of Stephen Bloodsworth in designing and producing the maps and quadrant charts.

Executive Summary

Introduction

The world is in the midst of a global technology revolution. For the past 30 years, advances in biotechnology, nanotechnology, materials technology, and information technology have been occurring at an accelerating pace, with the potential to bring about radical changes in all dimensions of life. The pace of these developments shows no sign of abating over the next 15 years, and it appears that their effects will be ever more remarkable. The technology of 2020 will integrate developments from multiple scientific disciplines in ways that could transform the quality of human life, extend the human lifespan, change the face of work and industry, and establish new economic and political powers on the global scene.

While people often do not understand a technology itself, they can often understand what that technology, when applied, might do for them and the societies in which they live when an application concept is presented to them. Actual adoption, however, is not necessarily automatic because of the confluence of economic, social, political, and other mitigating factors. Such technology applications, designed to accomplish specific functions, and their mitigating factors are the focus of our study.

Increasingly, such applications entail the integration of multiple technologies. New approaches to harnessing solar energy, for instance, are using plastics, biological materials, and nanoparticles. The latest water purification systems use nanoscale membranes together with biologically activated and catalytic materials. Technology applications such as these may help to address some of the most significant problems that different nations face—those involving water, food, health, economic development, the environment, and many other critical sectors.

While extensive, this technology revolution will play out differently around the globe. Although a technology application may be technically possible by 2020, not all countries will necessarily be able to acquire it—much less put it widely to use—within that time frame. An adequate level of science and technology (S&T) capacity is the first requirement for many sophisticated applications. A country might obtain a technology application through its domestic research and development (R&D) efforts, a technology transfer, or an international R&D collaboration—all various indicators of a country's S&T capacity. Or it could simply purchase a commercial off-the-shelf system from abroad. But many countries will not have achieved the necessary infrastructure or resources in 15 years to do such things across the breadth of the technology revolution.

What is more, the ability to acquire a technology application does not equal the ability to implement it. Doing research or importing know-how is a necessary initial step. But successful implementation also depends on the drivers within a country that encourage technological innovation and the barriers that stand in its way. Such drivers and barriers reflect a country's institutional, human, and physical capacity;[1] its financial resources; and its social, political, and cultural environment. Each of these factors plays a part in determining a nation's ability to put a new technology application into the hands of users, cause them to embrace it, and support its widespread use over time.

For these reasons, different countries will vary considerably in their ability to utilize technology applications to solve the problems they confront. To be sure, not all technology applications will require the same level of capacity to acquire and use. But even so, some countries will not be prepared in 15 years to exploit even the least demanding of these applications—even if they can acquire them—whereas other nations will be fully equipped to both obtain and implement the most demanding.

Some Top Technology Applications for 2020

Of 56 illustrative applications that we identified as possible by 2020, 16 appear to have the greatest combined likelihood of being widely available commercially, enjoying a significant market demand, and affecting multiple sectors (e.g., water, food, land, population, governance, social structure, energy, health, economic development, education, defense and conflict, and environment and pollution).

- *Cheap solar energy*: Solar energy systems inexpensive enough to be widely available to developing and undeveloped countries, as well as economically disadvantaged populations.
- *Rural wireless communications*: Widely available telephone and Internet connectivity without a wired network infrastructure.
- *Communication devices for ubiquitous information access*: Communication and storage devices—both wired and wireless—that provide agile access to information sources anywhere, anytime. Operating seamlessly across communication and data storage protocols, these devices will have growing capabilities to store not only text but also meta-text with layered contextual information, images, voice, music, video, and movies.
- *Genetically modified (GM) crops*: Genetically engineered foods with improved nutritional value (e.g., through added vitamins and micronutrients), increased production (e.g., by tailoring crops to local conditions), and reduced pesticide use (e.g., by increasing resistance to pests).
- *Rapid bioassays*: Tests that can be performed quickly, and sometimes simultaneously, to verify the presence or absence of specific biological substances.

---

[1] Institutional capacity includes honest and effective systems of governance, banking and finance, law, education, and health. Human capacity entails the quality and quantity of a country's educated and skilled personnel, as well as the level of education and scientific literacy of its people. Physical capacity involves the quality and quantity of critical infrastructures—e.g., transport and freight networks, schools, hospitals, research facilities, and utilities.

- *Filters and catalysts*: Techniques and devices to effectively and reliably filter, purify, and decontaminate water locally using unskilled labor.
- *Targeted drug delivery*: Drug therapies that preferentially attack specific tumors or pathogens without harming healthy tissues and cells.
- *Cheap autonomous housing*: Self-sufficient and affordable housing that provides shelter adaptable to local conditions, as well as energy for heating, cooling, and cooking.
- *Green manufacturing*: Redesigned manufacturing processes that either eliminate or greatly reduce waste streams and the need to use toxic materials.
- *Ubiquitous radio frequency identification (RFID) tagging of commercial products and individuals*: Widespread use of RFID tags to track retail products from manufacture through sale and beyond, as well as individuals and their movements.
- *Hybrid vehicles*: Automobiles available to the mass market with power systems that combine internal combustion and other power sources while recovering energy during braking.
- *Pervasive sensors*: Presence of sensors in most public areas and networks of sensor data to accomplish real-time surveillance.
- *Tissue engineering*: The design and engineering of living tissue for implantation and replacement.
- *Improved diagnostic and surgical methods*: Technologies that improve the precision of diagnoses and greatly increase the accuracy and efficacy of surgical procedures while reducing invasiveness and recovery time.
- *Wearable computers*: Computational devices embedded in clothing or in other wearable items, such as handbags, purses, or jewelry.
- *Quantum cryptography*: Quantum mechanical methods that encode information for secure transfer.

The technology applications we identified vary significantly in assessed technical feasibility and implementation feasibility by 2020. Table 1 shows the range of this variation on a matrix of 2020 technical feasibility versus 2020 implementation feasibility for all 56 technology applications. *Technical feasibility* is defined as the likelihood that the application will be possible on a commercial basis by 2020. *Implementation feasibility* is the net of all nontechnical barriers and enablers, such as market demand, cost, infrastructure, policies, and regulations. We based its assessment on rough qualitative estimates of the size of the market for the application in 2020 and whether or not it raises significant public policy issues. The numbers in parentheses are the number of sectors that the technology can affect, and the designation *global* (G) or *moderated* (M) indicates our estimate, based on both the technical foresights and our discussions with RAND regional experts, of whether the application will be diffused globally in 2020 or will be moderated in its diffusion (i.e., restricted by market, business sector, country, or region).

Table 1
Technical and Implementation Feasibility of Illustrative 2020 Technology Applications

		Implementation Feasibility			
		Niche market only (– –)	May satisfy a need for a medium or large market, but raises significant public policy issues (–)	Satisfies a well-documented need for a medium market and raises no significant public policy issues (+)	Satisfies a well-documented need for a large market and raises no significant public policy issues (++)
Technical Feasibility	Highly feasible (++)	• Chemical, biological, radiological, or nuclear (CBRN) sensors on emergency response teams (2,G)	• Genetic screening (2,G) • GM crops (8,M) • Pervasive sensors (4,G)	• Targeted drug delivery (5,M) • Ubiquitous information access (6,M) • Ubiquitous RFID tagging (4,G)	• Hybrid vehicles (2,G) • Internet (for purposes of comparison) (7,G) • Rapid bioassays (4,G) • Rural wireless communications (7,G)
	Feasible (+)	• GM animals for R&D (2,M) • Unconventional transport (5,M)	• Implants for tracking and identification (3,M) • Xenotransplantation (1,M)	• Cheap solar energy (10,M) • Drug development from screening (2,M) • Filters and catalysts (7,M) • Green manufacturing (6,M) • Monitoring and control for disease management (2,M) • Smart systems (1,M) • Tissue engineering (4,M)	• Improved diagnostic and surgical methods (2,G) • Quantum cryptography (2,G)
	Uncertain (U)	• Commercial unmanned aerial vehicles (6,M) • High-tech terrorism (3,M) • Military nanotechnologies (2,G) • Military robotics (2,G)	• Biometrics as sole identification (3,M) • CBRN sensor network in cities (4,M) • Gene therapy (2,G) • GM insects (5,M) • Hospital robotics (2,M) • Secure video monitoring (3,M) • Therapies based on stem cell R&D (5,M)	• Enhanced medical recovery (3,M) • Immunotherapy (2,M) • Improved treatments from data analysis (2,M) • Smart textiles (4,M) • Wearable computers (5,M)	• Electronic transactions (2,G) • Hands-free computer interface (2,G) • "In-silico" drug R&D (2,G) • Resistant textiles (2,G) • Secure data transfer (2,M)
	Unlikely (–)	• Memory enhancing drugs (3,M) • Robotic scientist (1,M) • "Super soldiers" (2,M)	• Chip implants for brain (4,M)	• Drugs tailored to genetics (2,M)	• Cheap autonomous housing (6,G) • Print-to-order books (2,G)
	Highly unlikely (– –)	• Proxy-bot (3,M) • Quantum computers (3,M)	• Genetic selection of offspring (2,M)	• Artificial muscles and tissue (2,M)	• Hydrogen vehicles (2,G)

Nations Will Continue to Vary in Their Capacity to Reap the Benefits of Technology Applications

Global diffusion of a technology application does not mean universal diffusion: Not every nation in the world will be able to implement, or even acquire, all technology applications by 2020. The level of direct S&T capacity may be markedly different from one country to another. Within different geographical regions, countries also have considerable differences that play into their ability. These differences may include variations in physical size, natural conditions (e.g., climate), and location (e.g., proximity to oceans and water). The size of the population and demographics (e.g., birthrate) may vary dramatically between countries in a single region. Countries may have very different types of government, economic systems, and levels of economic development.

The 29 countries we compared (Table 2) represent not only the world's major geographical regions but also the range of national differences within them. We selected many of these countries specifically as representative of groups of similar nations, trying not to include in a single geographical area more than one country with similar characteristics. If several countries in a given region were very large, for example, we brought in one that would grossly represent all the large countries. If a number of other nations in the same region were small, we included a representative small country.

What Countries Will Be Able to *Acquire* Which Technology Applications by 2020?

Seven of the 29 countries we compared will be *scientifically advanced* through 2020. They will almost certainly have the S&T capacity to acquire all 16 of the top technology applications by 2020. The United States and Canada in North America, Germany in Western Europe, and South Korea and Japan in Asia fall into this category. In Oceania, Australia takes its place on this list, as does Israel in the Middle East. These countries are in blue boxes in Figure 1.

Four of the 29 countries will be *scientifically proficient* through 2020. They will very likely have the necessary S&T capacity through 2020 to acquire 12 of the top 16 technology applications (see Figure 2). China and India in Asia, Poland in Eastern Europe and Russia represent this group. They are shown in green boxes in Figure 1.

Seven of the 29 countries will be *scientifically developing* through 2020. They will have sufficient S&T capacity through 2020 to acquire nine of the top 16 applications (see Figure 2).[2] From South America, Chile, Brazil, and Colombia fall into this group. Mexico in North

Table 2
Representative Countries Across Regions of the World Selected for Analysis

Asia	Oceania	North Africa and the Middle East	Europe	Africa	North America	Central and South America and the Caribbean
China	Australia	Egypt	Georgia	Cameroon	Canada	Brazil
India	Fiji	Iran	Germany	Chad	Mexico	Chile
Indonesia		Israel	Poland	Kenya	United States	Colombia
Japan		Jordan	Russia	South Africa		Dominican
South Korea			Turkey			Republic
Nepal						
Pakistan						

NOTE: We recognize that there are many ways to assign countries to regional groupings. In this instance, we placed Turkey in the European group because of the country's long and sustained commitment to join the European Union.

---

[2] Colombia will not be able to acquire ubiquitous RFID tagging because its economy is much less involved in international trade than the other countries in this group are, and its domestic and regional markets are unlikely to generate sufficient demand for this technology application

America, Turkey in Europe, Indonesia in Asia, and South Africa in Africa are also included. These seven countries are shown in yellow boxes in Figure 1.

Eleven of the 29 countries will be *scientifically lagging* through 2020. They will have only enough S&T capacity to acquire five of the applications through 2020 (see Figure 2). Fiji in Oceania; the Dominican Republic in the Caribbean; Georgia in Europe; Nepal and Pakistan in Asia; Egypt, Iran, and Jordan in North Africa and the Middle East; and Kenya, Cameroon, and Chad in Africa are in this group. These countries are shown in red boxes in Figure 1.

Figure 1
Selected Countries' Capacity to *Acquire* the Top 16 Technology Applications

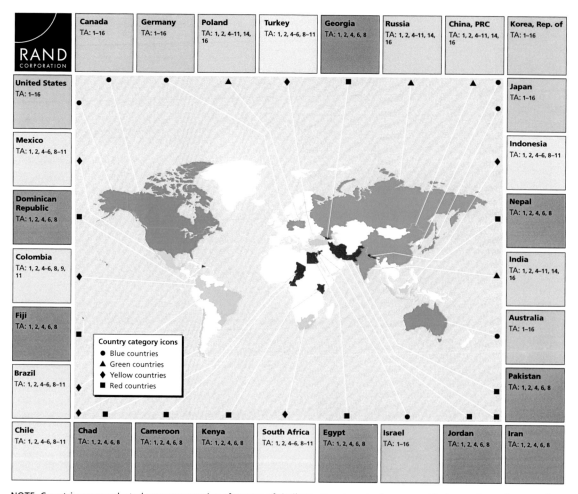

NOTE: Countries were selected as representative of groups of similar nations in a single geographical area. Countries are color coded by their S&T capacity: scientifically advanced (blue), scientifically proficient (green), scientifically developing (yellow), and scientifically lagging (red). Technology application (TA) numbers are as follows: (1) cheap solar energy, (2) rural wireless communications, (3) ubiquitous information access, (4) GM crops, (5) rapid bioassays, (6) filters and catalysts, (7) targeted drug delivery, (8) cheap autonomous housing, (9) green manufacturing, (10) ubiquitous RFID tagging, (11) hybrid vehicles, (12) pervasive sensors, (13) tissue engineering, (14) improved diagnostic and surgical methods, (15) wearable computers, (16) quantum cryptography.
RAND *MG475-1*

Figure 2
Mapping of Country Scientific Capability Rating to Top 16 Technology Applications

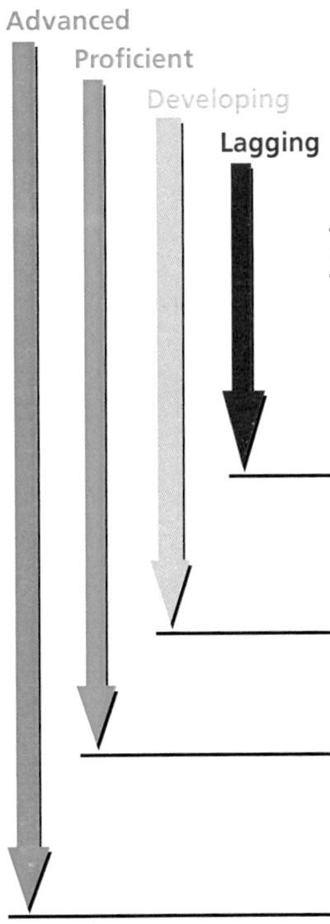

Needed Capability	Technology Applications
Low	Cheap solar energy Rural wireless communications GM crops Filters and catalysts Cheap autonomous housing
Medium	Rapid bioassays Green manufacturing Ubiquitous RFID tagging Hybrid vehicles
High	Targeted drug delivery Improved diagnostic and surgical methods Quantum cryptography
Very High	Ubiquitous information access Tissue engineering Pervasive sensors Wearable computers

Advanced · Proficient · Developing · Lagging

RAND *MG475-2*

By 2020, one should be able to see several trends in the capacity of countries to acquire technology applications (see Figure 1). Most of North America and Western Europe, along with Australia and the developed economies of East Asia, will be scientifically advanced. Most of Asia and Eastern Europe will be scientifically proficient. Latin America and much of Southeast Asia are likely to be scientifically developing. The majority of Africa and the Middle East, as well as the Caribbean and the Pacific Islands, will be scientifically lagging.

What Drivers and Barriers Affect These Countries' Ability to *Implement* the Technology Applications They Could Acquire?

The S&T capacity that enables a country to acquire a technology application is only one of several factors determining whether that country will be able to implement it. The *drivers* facilitating innovation and the *barriers* hindering it also have a decisive influence on the ability to *implement* technology applications (i.e., to put the applications in place and get significant gains from them across the country). These assessments involve such things as whom an application will benefit and whether a country can sustain its use over time. Drivers and barriers involve the same dimensions: A dimension that is a driver in one context may be a barrier in another. For example, financing, when available, would be a driver, but financing, when lacking, is a barrier. A high level of literacy among a nation's citizens would be a driver, but if literacy were low, it would form a barrier. And in certain cases, a dimension that is a barrier can simultaneously be a driver when only partial progress in that dimension has been made or when conflicting issues in the dimension are present. For example, education in the United States is a driver, but there are also concerns about problems in math and science education in the United States. Also, environmental concerns may dampen some S&T applications in China while promoting environmentally friendly applications, such as green manufacturing and hybrid vehicles.

These are the major drivers and barriers that countries may face through 2020 (see Figure 3):[3]

- *Cost and financing*: The cost of acquiring the technology application and of building the physical infrastructure and human capital to introduce and sustain its use, the mechanisms and resources available to access the needed funds, and the costs of those funds.
- *Laws and policies*: Legislation and policies that either promote, discourage, or prohibit the use of a particular technology application.
- *Social values, public opinion, and politics*: Religious beliefs, cultural customs, and social mores that affect how a technology application is perceived within a society; compatibility of a new application with dominant public opinions; and the politics and economics underlying debates about an application.
- *Infrastructure*: Physical infrastructure at a consistent threshold of quality that can be maintained, upgraded, and expanded over time.
- *Privacy concerns*: Social values toward privacy in a country and personal preferences about the availability and use of personal data that arise from an individual's ideological inclinations and experience with the privacy issue.
- *Use of resources and environmental health*: Availability and accessibility of natural resources, concerns about pollution and its impact on humans, and social attitudes and politics about conservation and preserving land and wildlife.
- *R&D investment*: Funding to educate and train scientists, engineers, and technicians; build research laboratories, computer networks, and other facilities; conduct scientific research and develop new technologies; transfer technologies to commercial applications; and enter technology applications into the marketplace.
- *Education and literacy*: Levels of general education and literacy adequate to make a population comfortable with technology and able to interface with it, and the availability of sufficiently high-quality postsecondary education and training in the sciences to stock a workforce comfortable with developing, using, and maintaining technology applications.
- *Population and demographics*: Overall size, average age, and growth rate of the population and the relative size of different age groups within a population.

---

[3] For a detailed discussion of the country driver and barrier assignments in Figure 3, see Silberglitt, Antón, Howell, and Wong (2006).

Figure 3
Drivers (D) and Barriers (B) in Selected Countries

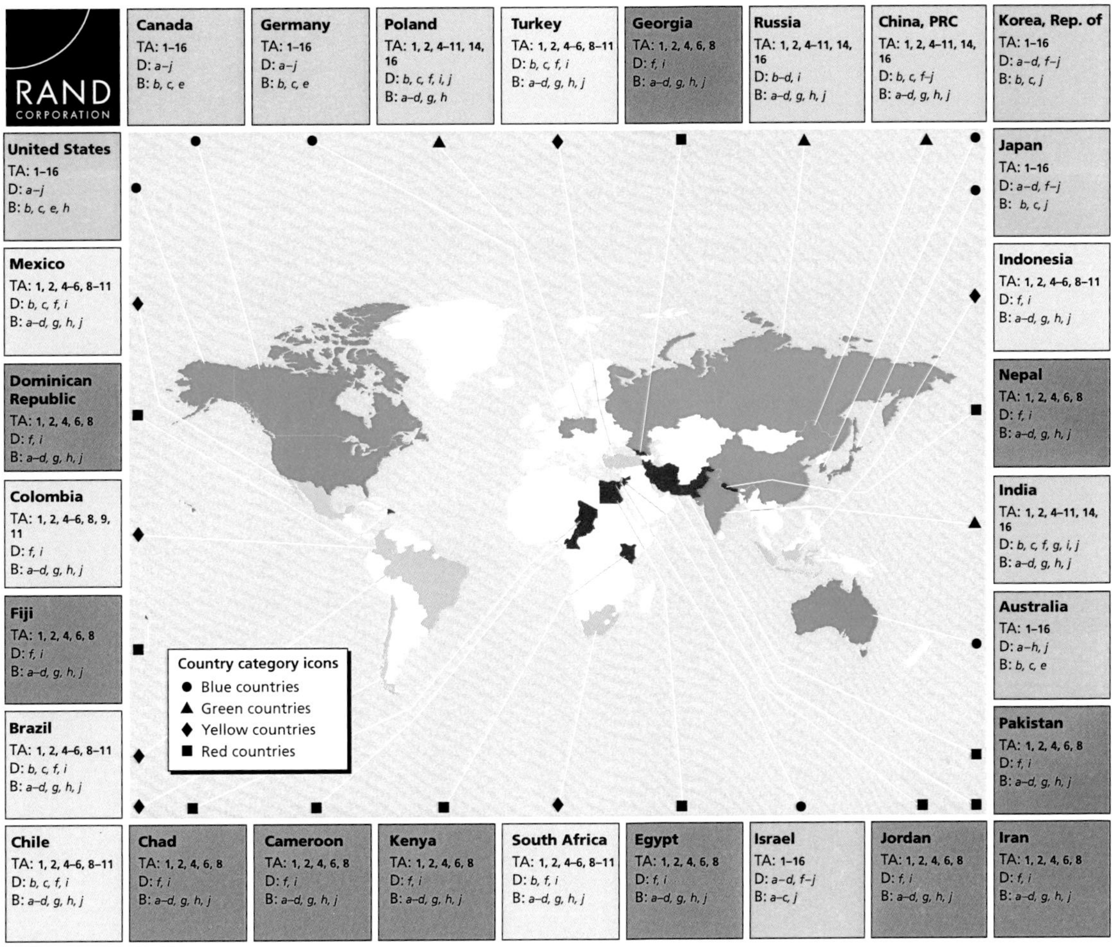

NOTE: Countries were selected as representative of groups of similar nations in a single geographical area. Countries are color coded by their S&T capacity: scientifically advanced (blue), scientifically proficient (green), scientifically developing (yellow), and scientifically lagging (red). Drivers (D) and barriers (B) are as follows: (a) cost and financing, (b) laws and policies, (c) social values, public opinion, and politics, (d) infrastructure, (e) privacy concerns, (f) use of resources and environmental health, (g) R&D investment, (h) education and literacy, (i) population and demographics, (j) governance and political stability. Technology application (TA) numbers are the same as in Figure 1: (1) cheap solar energy, (2) rural wireless communications, (3) ubiquitous information access, (4) GM crops, (5) rapid bioassays, (6) filters and catalysts, (7) targeted drug delivery, (8) cheap autonomous housing, (9) green manufacturing, (10) ubiquitous RFID tagging, (11) hybrid vehicles, (12) pervasive sensors, (13) tissue engineering, (14) improved diagnostic and surgical methods, (15) wearable computers, (16) quantum cryptography.
RAND MG475-3

- *Governance and political stability*: Degree of effectiveness or corruption within all levels of government; the influence of governance and stability on the business environment and economic performance; and the level of internal strife and violence, as well as external aggression; number and type of security threats.

Figure 4 illustrates the overall capacity of the 29 nations in our sample to implement all the technology applications they will be able to acquire.[4] Of the seven scientifically advanced countries able to obtain all 16 applications, the United States and Canada in North America and Germany in Western Europe will also be fully capable of implementing them through 2020. Japan and South Korea in Asia, Australia in Oceania, and Israel in the Middle East will be highly capable of implementing all 16 as well. All these countries will have excellent S&T capacity, along with the highest number of drivers and lowest number of barriers.

China will fall somewhat below these top seven countries; however, it will lead the group of scientifically proficient nations able to obtain 12 applications, with a high level of S&T capacity and many drivers. Still, because it will also possess numerous barriers, China will have to deal with more challenges to implementation than the group of scientifically advanced nations will. India, Poland, and Russia—the other three scientifically proficient countries—will be somewhat less capable than China of implementing the applications they can acquire. In these countries, although the S&T capacity will be high, the number of barriers will slightly exceed the number of drivers, making it more difficult to introduce and sustain the full range of possible technology applications.

All the countries in the scientifically developing group of nations—those able to acquire nine of 16 top applications—will have even less capacity than the proficient group will to implement them beyond laboratory research, demonstrations, or limited diffusion. Brazil and Chile in South America, Mexico in North America, and Turkey in Europe will be the most capable, followed by South Africa, then Indonesia, and finally Colombia. None of these seven countries will have a high level of S&T capacity. And each will have significantly more barriers than drivers.

The nations in the scientifically lagging group are able to obtain only five of the top 16 applications. Cameroon, Chad, and Kenya in Africa; the Dominican Republic in the Caribbean; Georgia in Europe; Fiji in Oceania; Egypt, Iran, and Jordan in North Africa and the Middle East; and Nepal and Pakistan in Asia will be the least capable of implementing these applications through 2020. With low levels of S&T capacity, these countries will also face numerous barriers and will benefit from very few drivers. It will therefore be very difficult for these countries to implement any but the simplest technology applications (see Figure 2).

---

[4] We analyzed country capacity to implement technology applications by taking into account three factors: (1) capacity to acquire, defined as the fraction of the top 16 technology applications listed for that country in Figure 1; (2) the fraction of the ten drivers for implementation applicable to that country; and (3) the fraction of the ten barriers to implementation applicable to that country. Figure 4 shows the position of each of the 29 representative countries on a plot for which the y-axis is the product of factors (1) and (2)—i.e., capacity to acquire scaled by the fraction of drivers—and the x-axis is factor (3). (Multiplying capacity to acquire by the fraction of drivers is consistent with the view that the absence of drivers reduces the probability that the technology applications a country can acquire will be implemented.) Both axes are shown as percentages: The y-axis starts at 0 percent (i.e., no capacity to acquire technology applications or drivers) and ends at 100 percent (i.e., capacity to acquire all 16 technology applications, with all 10 drivers applicable). The x-axis starts at 100 percent (i.e., all 10 barriers are applicable) and ends at 0 percent (i.e., no barriers are applicable). This figure provides a first-order assessment of the capacity to implement technology applications, in that we applied equal weighting to all technology applications, drivers, and barriers. We recognize that specific technology applications, drivers, and barriers might be more or less significant in particular countries.

Figure 4
Selected Countries' Capacity to *Implement* the Top 16 Technology Applications

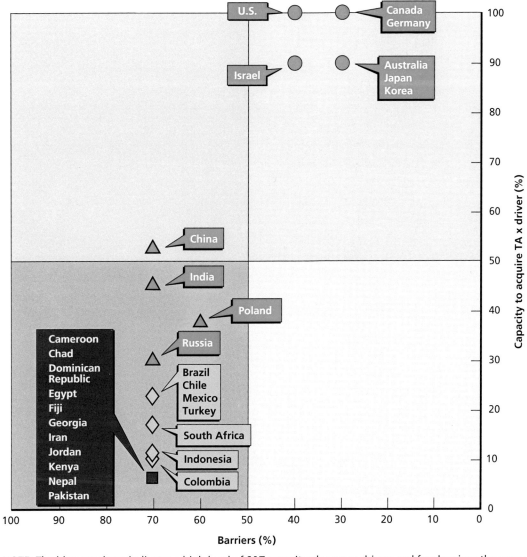

NOTE: The blue quadrant indicates a high level of S&T capacity plus many drivers and few barriers; the green quadrant indicates a high level of S&T capacity with many drivers and many barriers; the yellow quadrant indicates the lack of a high level of S&T capacity plus few drivers and few barriers; the red quadrant indicates the lack of a high level of S&T capacity with more barriers than drivers.
RAND *MG475-4*

None of the countries in our sample, regardless of their level of S&T capacity, will have low numbers of both drivers and barriers through 2020. This reflects the fact that nations cannot reduce barriers without simultaneously developing drivers and S&T resources.

The overall capacity of these representative nations to implement the technology applications they can acquire suggests the following trends:

- The technological preeminence of the scientifically advanced countries in North America, Western Europe, and Asia
- The emergence of China and India as rising technological powers, with the scientifically proficient countries of Eastern Europe, as represented by Poland, not far behind
- The relative slippage of Russia as a technological powerhouse
- The wide variation in technological capability among the scientifically developing countries of Southeast Asia and Latin America
- The large scientific and technological gap between the scientifically developing countries of Latin America, as well as Turkey and South Africa, and the rising technological powers, China and India
- The enormous scientific and technological gap between the scientifically lagging countries of Africa, the Middle East, and Oceania and the scientifically advanced nations of North America, Western Europe, and Asia
- The significant gap that must be filled before the scientifically lagging nations can even reach the level of proficiency.

Different Countries, Different Issues: The Capacity of Various Nations to Use Technology Applications to Address National Problems

The overall capacity of countries to implement the technology applications they can acquire provides a good general indication of the variation in how technology might change the world through 2020. It offers a comparative perspective on which countries are likely to be able to actually put technological opportunities into practice, which will be the technological powerhouses, which will be the emerging powers, and which will still be saddled with too many obstacles to benefit from the innovations of the next 15 years. It also suggests how much progress, in general, some countries need to make to exploit the technology revolution.

But just because a country has the capacity to implement a certain technology application does not mean that it will want or need to. With distinct sets of problems and diverse profiles, different countries will continue to have different national priorities through 2020. Because technology applications are designed to perform specific functions, they pertain only to certain problems. Consequently, not all 16 applications will be equally relevant for all countries. A country will be unlikely to invest in developing and implementing applications that will not help it achieve its most important goals.

The 16 top technology applications in our study can all help achieve at least several of the following objectives. In theory, all these goals will be important items on national agendas over the next 15 years:

- Promote rural economic development.
- Promote economic growth and international commerce.
- Improve public health.
- Improve individual health.

- Reduce the use of resources and improve environmental health.
- Strengthen the military and warfighters of the future.
- Strengthen homeland security and public safety.

Yet in practical terms, a country will give each of these objectives different priorities, depending on its state of economic and social development, internal politics, and domestic public opinion. Some countries may not even be in a position to pursue some of these goals because they have not yet achieved other, more fundamental, building blocks on which the goals rest. For example, promoting basic rural economic development may be a first step before pursuing international commerce.

Generally, a country's level of S&T capacity links up with indicators of economic and social development. By and large, countries with less S&T capacity also rank lower in the other two areas, while countries with more S&T capacity rank higher.[5] Consequently, nations with different levels of S&T capacity often share similar problems and, as a result, tend to prioritize similar objectives. Promoting rural economic development, improving public health, and reducing the use of resources and improving environmental health—all basic development goals—are usually top concerns for countries on the lowest rungs of the development ladder. More-developed countries may also give these goals prominence on their national agendas but often for different reasons and with less urgency. For example, scientifically developing countries are likely to be motivated to implement technology applications that can help them use resources more efficiently and clean up pollution mainly for the possible economic benefits, with environmental gains a secondary goal or spillover effect. As long as their economies are sluggish and living standards low, countries on this rung of the development ladder will not be in a position to pay up front for the long-term health gains of prioritizing environmental issues. Yet in countries whose economies are stronger and whose citizens can better afford (literally) to be concerned about the environment, public demand for cleaner, healthier surroundings and responsible stewardship of natural resources can drive the use of these applications.

Why Countries Prioritize Economic Growth

Economic growth and international commerce push nations up the development ladder. Consequently, promoting them becomes an increasingly important goal as countries build infrastructures, better educate their populations, and enter the global marketplace. For scientifically proficient countries, and even certain scientifically developing ones, driving economic growth can become a first-order concern. For scientifically advanced nations, this goal also usually takes top priority but for different reasons. The global marketplace is changeable and demanding, with new powers emerging and established ones continually vying for a competitive edge. To sustain current levels of prosperity and power, nations at the top of the develop-

---

[5] Compare the S&T capacity index in Appendix H of our in-depth analysis report (Silberglitt, Antón, Howell, and Wong, 2006) with the per capita gross domestic product and the Human Development Index rankings in Appendix J of the same document.

ment ladder must continually seek to push beyond what they already have. In this way, they can retain an advantage in the world of commerce and continue to improve the quality of life of their populations.

Countries at Various Levels of Development Prioritize Strengthening the Military

Strengthening a nation's military and warfighters does not necessarily or clearly correspond to a particular position on the development ladder. Certain countries sorely lacking in the most basic living standards have been observed to funnel the majority of their national budget into military spending, given certain circumstances. The same is true for strengthening homeland security and public safety. But in general, nations lower on the development ladder are not in a position to prioritize these two concerns. Meeting the essential needs of their populations—economic growth, health, nutrition, education, infrastructure—is their most urgent objective. Scientifically proficient and advanced countries with more power and more money can better afford to make these goals high priorities.

Individual Health as a National Priority Generally Follows Public Health

Improving individual health is by necessity a secondary goal for some nations. A country can usually only make this objective a matter of real national concern if its public health system is already functioning well and its population enjoys a high standard of living. For this reason, it is typically only a first-order goal in scientifically advanced countries. Technology applications that could help reduce infant mortality rates and increase the average life expectancy—both measures of good public health—are much more important for countries lower on the development ladder.

Countries' Capacity to Achieve Science and Technology Goals

Because national concerns tend to differ in these ways between countries with various levels of S&T capacity, particular sets of technology applications will be much more important, and their impacts much more dramatic, for certain nations than for others. But if a country were to establish a certain goal as a top priority in 2020 and resolve to address it, how capable would it be of actually implementing the applications that would enable it to do so? We looked at the scientifically lagging, developing, proficient, and advanced nations in our sample and for each one answered that question for the objectives likely to be relevant to countries at its level of S&T capacity.

Scientifically Lagging Countries

Countries in the scientifically lagging group tend to be at the bottom of the development ladder. Promoting rural economic development, improving public health, and reducing the use of resources and improving environmental health commonly rank highest on national agendas. The populations of many of the countries in our sample lack access to clean water and basic sanitation. Extreme poverty in rural areas can spur massive urban migration and discontent. Disease is often widespread. Essential resources, such as water and arable land, are frequently misused and rapidly dwindling. In many of these countries, the pervasive use of wood and coal-burning stoves is a major problem, generating indoor air pollution that has severe costs for the health of women and children in particular. The need for clean, cheap energy sources is urgent. With rapidly growing populations, low levels of literacy, and great disparities in wealth and power, these countries also frequently need to promote economic growth and international commerce. Stronger national economies would create jobs and generally improve the standard of living. But because very few countries at this level of S&T capacity are active participants in the global economy and because barriers are so abundant, this goal often takes a backseat to more basic development objectives.

All five of the technology applications these countries have the capacity to acquire—cheap solar energy, rural wireless communications, GM crops, filters and catalysts, and cheap autonomous housing—could help them both promote economic development in rural areas and improve public health. Solar energy would provide power for pumping water and irrigating crops, significantly improving agriculture and offering alternatives to subsistence farming (e.g., industrial cooperatives). It would also provide the power to run the filters that purify water supplies and the appliances to store medications. Better and more accessible water, food, and medicine would in turn improve public health. Providing lighting for homes and buildings and power for computers, solar energy could enable rural populations to participate in cottage industries and educate their children, growing the rural economy. Wireless communications would open the floodgate to economic development in remote areas, facilitating both commerce and education. Access to medical information and records would also significantly improve public health. GM crops would make food both more available and more nutritional, reducing the malnutrition and infant mortality that are so pervasive in these countries. Filters and catalysts would enable local populations to make unfit water sources usable and to clean wastewater for reuse. Cheap self-sufficient housing would provide rural populations with basic energy and shelter at minimal cost.

All five applications could also help these countries use fewer resources and improve environmental health. Cheap solar energy would provide energy without fuel combustion, reducing environmental emissions. Solar energy and cheap autonomous housing might help reduce the indoor air pollution generated by wood- and coal-burning stoves. Less reliance on firewood would promote healthy forests that would help control soil erosion; improve the quality of underground water; reduce sediment flows into rivers; and supply food, medicine, and construction materials. Rural wireless communications could help local and national governments monitor resources, environmental conditions, and pollution. GM crops would help conserve the natural resources used for agriculture and eliminate or reduce the magnitude of sources of pollution. Filters and catalysts would help conserve water and reduce waste streams.

Of the numerous technology applications that can promote economic growth in general, scientifically lagging nations will be able to acquire only two: cheap solar energy and rural wireless communications. Their benefits in this context lie mainly in helping to build more vital and productive rural economies that will be better able to contribute to their national economy and boost their global competitiveness.

If implemented broadly and sustainably, these technology applications have the potential to dramatically improve the quality of life of the vast majority of people living in poverty in these countries. But in practical terms, the nations in the scientifically lagging group will face considerable challenges in implementing any of the five—despite the fact that they place the least demand on S&T capacity. Drivers are scarce and barriers abundant in these countries. Unless the barriers are addressed, the lack of financial resources; institutional, physical, and human capacity; open markets; and transparent and stable governments will make it very difficult for the countries that could most benefit from these applications to put them to use.

Scientifically Developing Countries

Nations in the scientifically developing group commonly face many of the same problems as those in the scientifically lagging group. For example, in most of these countries, a significant percentage of the population is rural, with many people living at or below the poverty line. Outside the capital, infrastructure is typically poor. Provincial areas commonly lack cheap and stable electricity, a clean and dependable water supply, basic health services, good roads, and schools. As a result, urban populations in many of these nations are growing rapidly as people flock to the cities in hope of better economic opportunities. Consequently, promoting rural economic development is usually a top concern, to reduce rural poverty, soothe discontent, and slow urban migration.

Improving public health is often another leading goal. Because people in many of these countries frequently lack clean water and good sanitation, waterborne diseases are common and generally spread easily. The largely rural populations usually have little access to health care. In nations where cities are growing and people are traveling more frequently both domestically and abroad, the threat of epidemics can increase. In South Africa, for example, AIDS is taking a tremendous toll. Resources can present another major problem. In many nations at this level of S&T capacity, economic activities are further depleting already scarce natural resources and spoiling the environment. At the same time, energy prices are rising. For these reasons, it is imperative for many of them to use their resources more efficiently and improve the health of the environment.

Many of the countries with this level of S&T capacity frequently put promoting economic growth and international commerce higher on national agendas than scientifically lagging countries typically do (but still usually much lower than nations in the proficient and advanced groups). Most of them very much need to manage urban migration, create jobs, and expand the middle class. For countries that are to some degree actively exporting products to the global marketplace (e.g., Chile and Mexico), increasing economic competitiveness is a realistic development goal. Colombia is a clear exception in this regard; its economy is much less involved in international trade than most other nations in this group. The heightened political instability in some of the countries at this level could lead them to give increased prominence to strengthening homeland security and public safety. For example, in Colombia and Indonesia, political coups and military insurgencies are an ongoing threat.

Cheap solar energy, rural wireless communications, GM crops, filters and catalysts, and cheap autonomous housing could help scientifically developing nations promote economic development in rural areas, for the same reasons as in the scientifically lagging countries. These five, plus two others—rapid bioassays and green manufacturing—could help improve public health as well. The ability to use bioassays to quickly screen for diseases would enable governments to prevent epidemics. It would also increase the probability of correctly prescribing medications, decreasing resistance to antibiotics and other drugs. Reducing the volume of toxic materials in the environment produced by conventional manufacturing processes would improve public health.

Cheap solar energy, rural wireless communications, GM crops, filters and catalysts, green manufacturing, and hybrid vehicles could enable nations in this group to reduce the use of resources and improve environmental health. Again, the benefits would be the same as for the scientifically lagging countries. In addition, green manufacturing would diminish waste streams, allowing energy, water, and land to be used more efficiently; cut down pollutants in the environment; and reduce the burden on local governments of cleaning up polluted areas. Hybrid vehicles would significantly improve air quality, particularly in smog-ridden urban areas in these countries, where emissions are not tightly controlled. This problem is in part a result of urban migration. By addressing it, these countries would make it more appealing to move to the cities, which would allow the resulting economic growth without the negative environmental impact.

As in the lagging countries, cheap solar energy and rural wireless communications could help scientifically developing nations promote economic growth and international commerce. Rapid bioassays and ubiquitous RFID tagging, which nations at this level of S&T capacity can acquire as well, could be equally useful. Rapid bioassays would provide a means of ensuring that people can move safely across borders to conduct business, because it would allow governments to detect unintended transport of infectious disease more effectively. RFID tagging could enhance performance of retail industries, enabling greater control of inventories throughout the supply chain and making marketing more efficient.

Finally, for any country in this group that resolves to strengthen homeland security and public safety, rural wireless communications, rapid bioassays, filters and catalysts, and cheap autonomous housing could all help toward this end. Rural wireless communications would allow law enforcement and emergency response personnel to collect information from remote locations to prevent or respond to terror attacks, internal insurgencies, and disasters. Personnel on the scene would also be able to rapidly transfer information about the incident and response to local authorities. Rapid bioassays could help experts determine types of infections resulting from attacks, along with appropriate response measures. Filters and catalysts would provide potable water when water supplies are not safe. Cheap autonomous housing could provide temporary living quarters for relief workers and people made homeless by an incident.

Scientifically developing countries will vary significantly in their capacity to put technology applications into practice through 2020. Brazil, Chile, Mexico, and Turkey will be most capable of implementing relevant sets of applications (sometimes even on par with Russia in the proficient group). But compared with most of the proficient and advanced countries, their level of capacity will still be very low. South Africa will have even less capacity, and Colombia and Indonesia will have little more than that of the scientifically lagging countries. Overall, nations in this group will be most able to implement the applications that would spur the development of rural economies and reduce the use of resources. They will be somewhat less able to implement applications that could serve to improve public health. South Africa, Colombia, and Indonesia in particular may be severely impaired by the plethora of barriers they face. In terms of promoting economic growth, all the countries in this group will face considerable implementation challenges, and their capacity will be extremely low. These countries may develop more capacity if current positive economic and development trends continue, but without quality infrastructure beyond metropolitan areas, the use of relevant applications may be significantly limited. Finally, nations that aspire to strengthen homeland security will also have very limited capacity to implement the applications that can help in this area.

Scientifically Proficient Countries

Nations in the scientifically proficient group face a dynamic mix of problems. Promoting economic development and international commerce is often a top priority for countries with this level of S&T capacity but for very diverse reasons. The populations of China and India, for example, are quite large and continually growing. These countries urgently need to feed their many people, create jobs, and sustain wide-scale economic development. Yet while Poland and Russia have much smaller populations, economic growth is no less a concern. In the decade following the dissolution of the Soviet Union, Russia has encountered considerable economic difficulties. Although its population is shrinking in real terms, unemployment is high. The exodus of Russian scientists, engineers, and other professionals beginning in the 1990s has weakened the country's institutional and human capacity in science, health, and administration. Poland, as a relatively recent member of the European Union, is in a very different situation: It needs to bring its economy in line with EU standards.

In China and India, a significant fraction of the population is rural and impoverished. The rural economy is not much different from that of scientifically lagging and developing countries: Rapid economic growth is largely confined to urban areas, and rural and urban populations have great disparities in income, as well as health and education. In China in particular, the income gap is widening. Consequently, for both these nations, promoting rural economic development to reduce rural poverty is a much more pressing concern than it is for countries like Poland and Russia—although they still retain a national focus on promoting overall economic growth.

In many scientifically proficient countries, reducing the use of resources and improving environmental health is also among the most important objectives. Valuable assets such as arable land and fresh water—already scarce—are lost every day to land degradation, industrial pollution, and urban growth. In addition, many of these countries are at a level of development at which their populations are becoming increasingly aware of the high economic and health costs of environmental destruction and pollution.

For countries in the proficient group that lack clean water, electricity, and good sanitation in certain areas, improving public health is still a first-order concern. Countries at this level can suffer from the same public health issues as countries lower on the development ladder. Contagious diseases can spread easily, making epidemics a significant threat. Infant mortality rates can exceed international standards, and life expectancies can be lower than desirable. Yet at the same time, many countries with this level of S&T capacity are approaching the point on the development ladder where they can begin to aspire to improve individual health as well.

Strengthening the military and warfighters of the future is often a prominent concern for countries in the scientifically proficient group. For example, as a new EU member, Poland needs to modernize its military for greater compatibility with its new security partners. Russia wants to preserve its former status as a world military power. Strengthening homeland security and public safety can also be a relatively high priority. Russia, for instance, faces considerable internal security problems, such as organized crime and armed opposition in Chechnya.

As in the scientifically lagging and developing countries, cheap solar energy, rural wireless communications, rapid bioassays, and ubiquitous RFID tagging could promote economic growth and international commerce in the scientifically proficient countries. In addition, these countries will be able to acquire quantum cryptography, which, in providing a means of transferring information in a secure, reliable manner, could further aid economic development. This application would offer attractive benefits to banking and finance organizations, for example. Just as in the lagging and developing countries, cheap solar energy, rural wireless communications, GM crops, filters and catalysts, and cheap autonomous housing could enable those scientifically proficient nations that make it a priority to do so to promote rural economic development.

In terms of improving public health, the same applications that the developing countries have the S&T capacity to acquire toward this end—cheap solar energy, rural wireless communications, GM crops, filters and catalysts, cheap autonomous housing, rapid bioassays, and green manufacturing—could help the proficient nations as well. In addition, these countries have the S&T capacity to acquire targeted drug delivery, which is likely to eventually become such a widespread application that it will enable cancers and other diseases to be treated on site in remote areas, with significant benefits to public health. Similarly, they will be able to acquire the same applications the developing countries to reduce the use of resources and improve environmental health: cheap solar energy, rural wireless communications, GM crops, filters and catalysts, green manufacturing, and hybrid vehicles.

The benefits to public health from cheap solar energy, rural wireless communications, GM crops, rapid bioassays, filters and catalysts, cheap autonomous housing, and green manufacturing would also better the health of individuals. In addition, targeted drug delivery, by limiting damage to healthy cells and tissues when administering therapies, would enable less-invasive, debilitating treatments and better outcomes. Improved diagnostic and surgical methods would make diagnoses more precise and surgical procedures more effective, and reduced recovery times would give a wider group of patients the option of surgery.

Rural wireless communications, rapid bioassays, filters and catalysts, cheap autonomous housing, ubiquitous RFID tagging, and quantum cryptography would help these proficient nations strengthen their military and warfighters. Military command, control, and communication could be improved with rural wireless communications. Rapid bioassays would allow military medical personnel to identify weapon-grade pathogens in the environment. Filters and catalysts could be employed in situations involving chemical or biological contaminants. Cheap autonomous housing could provide personnel on the ground with improved living quarters. RFID tagging would enable command centers to track the location and conditions of personnel engaged in operations. Quantum cryptography could safeguard tactical communications.

These technology applications could also enhance homeland security and public safety. The benefits would be the same as for the scientifically developing countries. Quantum cryptography could protect critical data and networks from hackers and attackers. In addition, targeted drug delivery, also obtainable by the proficient nations, could expedite responses to chemical and biological attacks and minimize casualties.

In terms of capacity to implement, China consistently has the most, followed by India, and then Poland. In every case, Russia trails, with the least capacity in the group to implement the relevant applications for any of the problem areas. As a whole, these countries have a fairly high capacity to put applications into practice to promote rural economic development and to reduce the use of resources and improve environmental health. Their ability to improve public health will be only slightly less than that. In the first two cases, China approaches the capacity level of several of the scientifically advanced countries, with India not far behind. Russia, in contrast, has no more capacity than the most capable of the scientifically developing nations. The scientifically proficient countries will be moderately capable of implementing the applications that would improve individual health. Implementation capacity will still be substantial but somewhat less for strengthening the military and warfighters and increasing homeland security and public safety. As much as these countries may need to achieve this goal, promoting economic growth and international commerce will be the most challenging of all. The capacity of the proficient countries to implement the relevant applications toward this end will be less than for all the other goals. There will be a very large gap, for example, between their ability to use technology applications to develop their international economy and that to improve public health or reduce the use of resources.

Scientifically Advanced Countries

Nations with the highest level of S&T capacity sit atop the development ladder. Their leading concerns are usually quite different from those of countries with less capacity because they have already achieved the more basic development objectives prerequisite to focusing on those goals. When a national priority is the same, a scientifically advanced country often has very different motivations from those of a lagging or developing one. Promoting economic growth and international commerce is a case in point. The nations in this group are already world economic leaders; their problem is usually to maintain or capture even more of a competitive advantage in an aggressive global market. South Korea, for example, has to deal with a China rapidly gaining S&T capacity and emerging as a commanding economic force. It also needs to gain ground on Japan, the United States, and other economic superpowers. Other advanced countries are contending with skyrocketing health costs. With rapidly aging populations, they need to increase the productivity of their future workforce to finance cutting-edge medical treatment.

Aging populations and a high standard of living also put improving individual health at the head of the national agenda in many scientifically advanced countries. Enhancing public health is often an objective, too, but usually a much less prominent one, given that these nations have already achieved very effective public health systems and will gain only marginal benefits. Exceptional circumstances, such as a need to provide emergency medical relief should a disaster strike, usually drive this goal.

Energy can be very costly in some countries in this group. At the same time, public awareness of the negative impacts of pollution and inefficient management of resources is often high. Consequently, citizens in nations at this level of S&T capacity frequently demand cleaner environments and more responsible consumption of natural assets. This can make reducing the use of resources and improving environmental health an important national objective.

Strengthening homeland security and public safety is a principal concern for some nations at this level of S&T capacity. While some nations have had terrorism prevention on their national agendas for a long time, this issue has become more prominent as a number of advanced countries have had recent experiences with terrorism—the United States with the attacks of September 11, 2001, and Spain and the United Kingdom with train bombings, for instance. Public demand in such countries to reduce internal security threats can run very high. Making the military and warfighters of the future stronger is often among their foremost concerns as well, for varying reasons. Both Israel and South Korea face potential threats from hostile neighboring countries; the United States seeks to maintain its global military predominance.

Just as for countries with less S&T capacity that can acquire these applications, cheap solar energy, rural wireless communications, rapid bioassays, ubiquitous RFID tagging, and quantum cryptography could also help the scientifically advanced nations promote economic growth and international commerce. But these countries will be able to acquire more sophisticated applications as well—ubiquitous information access, pervasive sensors, tissue engineering, and wearable computers. Agile access to information could improve productivity, create new avenues for conducting business on the run, and expand global Internet commerce. Pervasive sensors could help manage logistics, determine market demand, and safeguard electronic transactions. Expertise in sensor development and data management would expand a company's commercial opportunities. The technical or medical expertise to engineer tissue, the capability to manufacture it, or any related intellectual property rights would have the same effect. Wearable computers would open exciting new doors for economic sectors based on computation.

To improve individual health, the scientifically advanced nations could acquire cheap solar energy, rural wireless communications, GM crops, rapid bioassays, filters and catalysts, targeted drug delivery, cheap autonomous housing, green manufacturing, tissue engineering, and improved diagnostic and surgical methods. In addition, ubiquitous information access would make health information available anywhere and anytime and facilitate information sharing between patients and providers. Tissue engineering would minimize medical complications and recurrences by providing new ways of treating wounds, disease, and injuries. It might also permit classes of chronically ill or formerly untreatable individuals to join the workforce. Wearable computers could enable patients or their doctors to continuously monitor patients' health status. Along with the relevant applications obtainable by countries lower on the development ladder, ubiquitous information access would also contribute to improving public health at this level of S&T capacity.

All the applications that could help reduce the use of resources and improve environmental health would be available to the advanced nations: cheap solar energy, rural wireless communications, GM crops, filters and catalysts, green manufacturing, and hybrid vehicles. To strengthen homeland security and public safety, advanced countries will be able to acquire rural wireless communications, rapid bioassays, filters and catalysts, targeted drug delivery, cheap autonomous housing, and quantum cryptography. In addition, ubiquitous access to information would facilitate information sharing and increase the ability to track individual's activities. Pervasive sensors would provide governments with a powerful tool for law enforcement. Together with miniaturized communications devices, wearable computers could enable personnel to send and receive instructions in conflict situations.

Rural wireless communications, rapid bioassays, filters and catalysts, cheap autonomous housing, ubiquitous RFID tagging, and quantum cryptography could all help strengthen the military and warfighters. Beyond these, ubiquitous information access could improve combat planning and execution, logistics, and support functions. Pervasive sensors could be implemented in tactical situations to provide updated intelligence and targeting. The increased ability to exchange instructions provided by wearable computers would be a significant advantage in military situations as well.

For any of the national objectives that they choose to prioritize, all the scientifically advanced countries will be highly capable of implementing the full set of relevant technology applications. With abundant drivers, relatively few barriers, and unrivaled S&T ability, these advanced countries are the only ones among our sample likely to be able to implement, on a broad scale, the applications that demand the highest level of infrastructure and institutional, physical, and human capacity.

The Science and Technology Path to 2020

As the global technology revolution proceeds over the next 15 years, it will follow a trajectory with certain defining characteristics.

Accelerated Technology Development Will Continue

We see no indication that the rapid pace of technology development will slow in the next decade and a half. Neither will the trends toward multidisciplinarity and the increasingly integrated nature of technology applications reverse. Indeed, most of the top 16 technology applications for 2020 draw from at least three of the areas addressed in this study—biotechnology, nanotechnology, materials technology, and information technology—and many involve all four. Underlying these trends are global communications (Internet connectivity, scientific conferences, and publications) and instrumentation advances (the development and cross-fertilization of ever more-sensitive and selective instrumentation).

Countries Will Benefit in Considerably Different Ways

Over the next 15 years, certain countries will possess vastly different S&T capacities. They will also vary considerably in the institutional, human, and physical capacity required to develop drivers for implementing technology applications and overcome barriers. Consequently, the global technology revolution will play out quite differently among nations.

The scientifically advanced countries of North America, Western Europe, and Asia, along with Australia, are likely to gain the most, as exemplified by their capacity to acquire and implement all the top 16 example technology applications. For whatever problems and issues that rank high on their national agendas, they will be able to put into practice a wide range of applications to help address them.

If they can address multiple barriers to implementation, emerging economies, such as China and India in Asia and Brazil and Chile in South America, will be able to use technology applications to support continued economic growth and human development for their populations. Emerging technological powers China and India will have the best opportunity to approach the ability of the scientifically advanced countries to use applications to achieve national goals. The scientifically proficient countries of Eastern Europe, as represented by Poland, appear to be poised next in line behind China and India. In contrast, it looks likely that Russia's capacity to implement technology applications will continue to deteriorate, with the most advanced of the scientifically developing countries (represented by Brazil, Chile, Mexico, and Turkey) potentially overtaking her.

The scientifically lagging countries around the world will face the most severe problems—disease, lack of clean water and sanitation, and environmental degradation. They will also likely lack the resources to address these problems. Consequently, they stand to gain the most from implementing the 2020 technology applications. However, to do so, these nations will need to make substantial inroads in building institutional, physical, and human capacity. The efforts and sponsorship of international aid agencies and other countries may assist in these efforts, but the countries themselves will have to improve governance and achieve greater stability before they will be able to benefit from available S&T innovations.

Action Will Be Required to Maintain a High Level of S&T Capacity

The accelerating pace of technology development and the growing capacity of emerging economies to acquire and implement technology applications will make economic security a moving target even for the most advanced nations. If countries are to stay ahead in their capacity to implement applications, they will need to make continuing efforts to ensure that laws, public opinion, investment in R&D, and education and literacy are drivers for, and not barriers to, technology implementation. In addition, they will have to build and maintain whatever infrastructure is needed to implement the applications that will give them a competitive advantage.

Countries That Lack Capacity Will Need to Build It

For scientifically lagging and developing countries, implementing technology applications to address problems and issues will not be primarily about technology, or even S&T capacity. The greater challenge they will face is the lack of institutional, human, and physical capacity, including effective and honest governance. Development results from improvements in economic growth, social equity, health and the environment, public safety and security, and good governance and stability. The countries with the best performance in these indicators of development will most likely have the greatest institutional, human, and physical capacity to implement technology applications. Less-developed countries that hope to benefit from technology applications will have to improve their performance in these development areas to build the requisite institutional, human, and physical capacity.

Certain Technology Applications Will Spark Heated Public Debate

Several of the top 16 technology applications will raise significant public policy issues that will trigger strong, and sometimes conflicting, reactions and opinions between countries, regions, and ethnic, religious, cultural, and other interest groups. Many of the most controversial applications will involve biotechnology (e.g., GM crops). Others, such as pervasive sensors and certain uses of RFID implants to track and identify people, will potentially have provocative implications for personal privacy and freedom. Yet any controversy that flares up will probably not be the same around the world. A technology application that raises extremely divisive questions in one country may cause no stir at all in another because of different social values.

Consideration Could Head Off Problems and Maximize Benefits

Public policy issues will need to be resolved before a country will be able to realize the full benefits of a technology application. Not all technology may be good or appropriate in every circumstance, and just because a country has the capacity to implement a technology application does not necessarily mean that it should. Ethical, safety, and public concerns will require careful analysis and consideration. Public policy issues will need to be debated in an environment that seeks to resolve conflicts. Such public debate, in addition to being based on sound data, will need to be inclusive and sensitive to the range of traditions, values, and cultures within a society. In some cases, issues will remain after the debate, slowing or even stopping technology implementation. Sometimes the reasons clearly will be good (e.g., when safety concerns cannot be adequately addressed), and sometimes the result will simply reflect collective decisionmaking determining what a particular society wants and does not want.

A Few Words in Conclusion

As the global technology revolution proceeds, market forces will moderate and vector its course, its technology applications, and their implementation. Predicting the net effect of these forces is predicting the future—wrought with all the difficulties of such predictions. But current technology trends have substantial momentum behind them and will certainly be the focus of continued R&D, consideration, and debate over the next 15 years. By 2020, countries will be applying many of these technologies in some guise or other and the effects will be significant, changing lives across the globe.

Selected Bibliography

For a more detailed discussion of the material described in this report, including further documentation and references, the reader is strongly encouraged to review our in-depth analyses in the following companion report:

Silberglitt, Richard, Philip S. Antón, David R. Howell, and Anny Wong, with Natalie Rose Gassman, Brian A. Jackson, Eric Landree, Shari Lawrence Pfleeger, Elaine M. Newton, and Felicia Wu, *The Global Technology Revolution 2020—In-Depth Analyses: Bio/Nano/Materials/Info Trends, Drivers, Barriers, and Social Implications*, Santa Monica, Calif.: RAND Corporation, TR-303-NIC, 2006. Online at http://www.rand.org/pubs/technical_reports/TR303/index.html.

Technology Foresight

Applewhite, Ashton, "The View from the Top: Forty Leading Lights Ponder Tech's Past and Consider Its Future," *IEEE Spectrum,* Vol. 41, No. 11, November 2004, pp. 36–51.

Christensen, Clayton M., *The Innovator's Dilemma*, New York: HarperCollins, 2003.

Glenn, Jerome C., and Theodore J. Gordon, *Future S&T Management Policy Issues: 2025 Global Scenarios*, Washington, D.C.: Millennium Project, American Council for the United Nations University, undated. Online at http://www.acunu.org/millennium/scenarios/st-scenarios.html (as of March 2006).

———, "Millennium 3000 Scenarios," excerpt from *State of the Future at the Millennium*, Washington, D.C.: Millennium Project, American Council for the United Nations University, undated. Online at http://www.acunu.org/millennium/m3000-scenarios.html (as of March 2006).

———, *2004 State of the Future,* executive summary, Washington, D.C.: Millennium Project, American Council for the United Nations University, 2004.

National Intelligence Council, *Global Trends 2015: A Dialogue About the Future with Nongovernment Experts,* Central Intelligence Agency, NIC 2000-02, 2000. Online at http://www.cia.gov/cia/reports/globaltrends2015/ (as of March 2006).

———, *Mapping the Global Future: Report of the National Intelligence Council's 2020 Project,* Central Intelligence Agency, NIC 2004-13, 2004. Online at http://www.cia.gov/nic/NIC_globaltrend2020.html (as of March 2006).

Salo, Ahti, and Kerstin Cuhls, "Technology Foresight: Past and Future," *Journal of Forecasting*, Vol. 22, Nos. 2–3, March–April 2003, pp. 79–82.

Technology Trends and Applications

Anderson, Robert H., Philip S. Antón, Steven C. Bankes, Tora K. Bikson, Jonathan P. Caulkins, Peter J. Denning, James A. Dewar, Richard O. Hundley, and C. Richard Neu, *The Global Course of the Information Revolution: Technological Trends—Proceedings of an International Conference*, Santa Monica, Calif.: RAND Corporation, CF-157-NIC, 2000. Online at http://www.rand.org/pubs/ conf_proceedings/CF157/index.html (as of March 2006).

Antón, Philip S., Richard Silberglitt, and James Schneider, *The Global Technology Revolution: Bio/ Nano/Materials Trends and Their Synergies with Information Technology by 2015*, Santa Monica, Calif.: RAND Corporation, MR-1307-NIC, 2001. Online at http://www.rand.org/pubs/ monograph_reports/MR1307/index.html (as of March 2006).

Cohen, Smadar, and Jonathan Leor, "Rebuilding Broken Hearts," *Scientific American*, Vol. 291, No. 5, November 2004, pp. 44–51. Online at http://www.sciam.com/article.cfm?articleID= 0007428C-6DD4-1178-AD6883414B7F0000 (as of March 2005).

Hundley, Richard O., Robert H. Anderson, Tora K. Bikson, and C. Richard Neu, *The Global Course of the Information Revolution: Recurring Themes and Regional Variations*, Santa Monica, Calif.: RAND Corporation, MR-1680-NIC, 2003. Online at http://www.rand.org/pubs/monograph_reports/ MR1680/index.html (as of March 2006).

International Roadmap Committee, "Emerging Research Devices," in *International Technology Roadmap for Semiconductors, 2003 Edition*, San Jose, Calif.: Semiconductor Industry Association, 2003. Online at http://public.itrs.net/Files/2003ITRS/Home2003.htm (as of March 2006).

———, *International Technology Roadmap for Semiconductors: 2004 Update*, San Jose, Calif.: Semiconductor Industry Association, 2004. Online at http://www.itrs.net/Common/2004Update/ 2004Update.htm (as of March 2006).

Langer, Robert, and David A. Tirrell, "Designing Materials for Biology and Medicine," *Nature*, Vol. 428, No. 6982, April 1, 2004, pp. 487–492.

National Nanotechnology Initiative, "What Is Nanotechnology?" Web page, undated, http://www. nano.gov/html/facts/whatIsNano.html (as of March 2006).

Royal Academy of Engineering, *Nanoscience and Nanotechnologies: Opportunities and Uncertainties*, London, 2004. Online at http://www.raeng.org.uk/policy/reports/nanoscience.htm (as of March 2006).

U.S. Department of Commerce, Technology Administration, *A Survey of the Use of Biotechnology in Industry*, October 2003. Online at http://www.technology.gov/reports/Biotechnology/CD120a_ 0310.pdf (as of March 2006).

U.S. National Science Foundation and U.S. Department of Commerce, *Converging Technologies for Improving Human Performance: Nanotechnology, Biotechnology, Information Technology and Cognitive Science*, Arlington, Va., June 2002. Online at http://www.wtec.org/ConvergingTechnologies (as of March 2006).

Country Capacity to Acquire and Implement Technology Applications

Bhargava, Vinay, and Emil Bolongaita, *Challenging Corruption in Asia: Case Studies and a Framework for Action*, Washington, D.C.: The World Bank, 2004.

Central Intelligence Agency, *The World Factbook*, July 14, 2005. Online at http://www.cia.gov/cia/publications/factbook/ (as of March 2006).

Freedom House, *Freedom in the World 2005: The Annual Survey of Political Rights and Civil Liberties*, 2004. Online at http://www.freedomhouse.org/template.cfm?page=15&year=2005 (as of March 2006).

Kaufman, Daniel, and Aart Kraay, "Governance and Growth: Causality Which Way?—Evidence for the World, in Brief," The World Bank, February 2003. Online at http://www.worldbank.org/wbi/governance/pdf/growthgov_synth.pdf (as of March 2006).

Kaufman, Daniel, Aart Kraay, and Massimo Mastruzzi, *Governance Matters IV: Governance Indicators for 1996–2004*, Washington, D.C.: World Bank Institute, May 9, 2005. Online at http://www.worldbank.org/wbi/governance/pubs/govmatters4.html (as of March 2006).

Transparency International, "Transparency International Global Corruption Barometer," July 3, 2003. Online at http://www.transparency.org/policy_and_research/surveys_indices (as of March 2006).

———, Corruption Perception Index 2005. Online at http://ww1.transparency.org/cpi/2005/cpi2005_infocus.html (as of March 2006).

Wagner, Caroline, Irene Brahmakulam, Brian Jackson, Anny Wong, and Tatsuro Yoda, *Science and Technology Collaboration: Building Capacity in Developing Countries*, Santa Monica, Calif.: RAND Corporation, MR-1357.0-WB, 2001. Online at http://www.rand.org/pubs/monograph_reports/MR1357.0/index.html (as of March 2006).

Wilson, Ernest J., *The Information Revolution and Developing Countries*, Cambridge, Mass.: MIT Press, 2004.

Wong, Anny, Aruna Balakrishnan, James Garulski, Thor Hogan, Eric Landree, and Maureen McArthur, *Science and Technology Research and Development Capacity in Japan: Observations from Leading U.S. Researchers and Scientists*, Santa Monica, Calif.: RAND Corporation, TR-211-MRI, 2004. Online at http://www.rand.org/pubs/technical_reports/TR211/index.html (as of March 2006).

Problems and Issues

Board on Global Health and Institute of Medicine, *Microbial Threats to Health: Emergence, Detection, and Response*, Washington, D.C.: National Academies Press, 2003. Online at http://www.nap.edu/books/030908864X/html/ (as of March 2006).

Cohen, Desmond, "Poverty and HIV/AIDS in Sub-Saharan Africa," Issue Paper No. 27, United Nations Development Programme, undated. Online at http://www.undp.org/hiv/publications/issues/english/issue27e.html (as of March 2006).

Food and Agriculture Organization of the United Nations, *The State of Food and Agriculture, 2003–2004—Agricultural Biotechnology: Meeting the Needs of the Poor?* Rome, Italy: United Nations, 2004. Online at http://www.fao.org/documents/show_cdr.asp?url_file=/docrep/006/Y5160E/Y5160E00.htm (as of March 2006).

National Research Council, *Review of the Research Program of the Partnership for a New Generation of Vehicles: Seventh Report*, Washington, D.C.: National Academy Press, 2001.

———, *The Hydrogen Economy Opportunities, Costs, Barriers, and R&D Needs*, Washington, D.C.: National Academies Press, 2004.

Pew Initiative on Food and Biotechnology, *Bugs in the System? Issues in the Science and Regulation of Genetically Modified Insects*, Washington, D.C.: Pew Charitable Trusts, 2004. Online at http://www.pewtrusts.com/pdf/pifb_bugs_012204.pdf (as of March 2006).

Sustainable Mobility Project, *Mobility 2030: Meeting the Challenges to Sustainability*, Geneva, Switzerland: World Business Council for Sustainable Development, 2004. Online at http://www.wbcsd.org/Plugins/DocSearch/details.asp?DocTypeId=25&ObjectId=NjA5NA&URLBack=%2Ftemplates%2FTemplateWBCSD2%2Flayout%2Easp%3Ftype%3Dp%26MenuId%3DMjYz%26doOpen%3D1%26ClickMenu%3DLeftMenu (as of March 2006).

U.S. Department of Energy, "Basic Research Needs for the Hydrogen Economy," Basic Energy Sciences Workshop on Hydrogen Production, Storage, and Use, Rockville, Md., May 13–15, 2003. Online at http://www.sc.doe.gov/bes/reports/abstracts.html#NHE (as of March 2006).

Woodward, John D., Katharine Watkins Webb, Elaine M. Newton, Melissa Bradley, David Rubenson, Kristina Larson, Jacob Lilly, Katie Smythe, Brian K. Houghton, Harold Alan Pincus, Jonathan M. Schachter, and Paul Steinberg, *Army Biometric Applications: Identifying and Addressing Sociocultural Concerns*, Santa Monica, Calif.: RAND Corporation, MR-1237-A, 2001. Online at http://www.rand.org/pubs/monograph_reports/MR1237/index.html (as of March 2006).

World Health Organization, "Water-Related Diseases," Web page, undated, http://www.who.int/water_sanitation_health/diseases/oncho/en/ (as of March 2006).

———, *Emerging Issues in Water and Infectious Diseases*, Geneva, Switzerland, 2003. Online at http://www.who.int/water_sanitation_health/emerging/emergingissues/en/ (as of March 2006).

———, *Avian Influenza: Assessing the Pandemic Threat*, Geneva, Switzerland, 2005.

———, Food Safety Department, *Modern Food Biotechnology, Human Health and Development*, Geneva, Switzerland, 2005. Online at http://www.who.int/foodsafety/biotech/who_study/en/ (as of March 2006).

Wu, Felicia, and William P. Butz, *The Future of Genetically Modified Crops: Lessons from the Green Revolution*, Santa Monica, Calif.: RAND Corporation, MG-161-RC, 2004. Online at http://www.rand.org/pubs/monographs/MG161/index.html (as of March 2006).